Rough Fuzzy
Image Analysis

Foundations and Methodologies

CHAPMAN & HALL/CRC MATHEMATICAL AND COMPUTATIONAL IMAGING SCIENCES

Series Editors

Chandrajit Bajaj

Center for Computational Visualization
The University of Texas at Austin

Guillermo Sapiro

Department of Electrical
and Computer Engineering
University of Minnesota

Aims and Scope

This series aims to capture new developments and summarize what is known over the whole spectrum of mathematical and computational imaging sciences. It seeks to encourage the integration of mathematical, statistical and computational methods in image acquisition and processing by publishing a broad range of textbooks, reference works and handbooks. The titles included in the series are meant to appeal to students, researchers and professionals in the mathematical, statistical and computational sciences, application areas, as well as interdisciplinary researchers involved in the field. The inclusion of concrete examples and applications, and programming code and examples, is highly encouraged.

Proposals for the series should be submitted to the series editors above or directly to:
CRC Press, Taylor & Francis Group
4th, Floor, Albert House
1-4 Singer Street
London EC2A 4BQ
UK

CHAPMAN & HALL/CRC
MATHEMATICAL AND COMPUTATIONAL IMAGING SCIENCES

Rough Fuzzy Image Analysis

Foundations and Methodologies

Edited by

Sankar K. Pal

James F. Peters

CRC Press
Taylor & Francis Group
Boca Raton   London   New York

CRC Press is an imprint of the
Taylor & Francis Group, an **informa** business

A CHAPMAN & HALL BOOK

CRC Press
Taylor & Francis Group
6000 Broken Sound Parkway NW, Suite 300
Boca Raton, FL 33487-2742

First issued in paperback 2017

ISBN 13: 978-1-138-11623-8 (pbk)
ISBN 13: 978-1-4398-0329-5 (hbk)

Library of Congress Cataloging-in-Publication Data

Rough fuzzy image analysis : foundations and methodologies / editors, Sankar K. Pal, James F. Peters.
    p. cm.
    "A CRC title."
    Includes bibliographical references and index.
    ISBN 978-1-4398-0329-5 (hardcover : alk. paper)
    1. Image analysis. 2. Fuzzy sets. I. Pal, Sankar K. II. Peters, James F. III. Title.

TA1637.R68 2010
621.36'7--dc22   2009053741

Visit the Taylor & Francis Web site at
http://www.taylorandfrancis.com

and the CRC Press Web site at
http://www.crcpress.com

Printed and bound by CPI Group (UK) Ltd, Croydon, CR0 4YY

21/10/2024

01777046-0005